Table of Contents

Contents

1963 Events

May 1-8 Civil Rights protestors march in Birmingham

June 11 Thich Quang Duc, a Buddhist monk, burns himself to death on a Saigon street to protest persecution of Buddhists by the South Vietnamese government, on.

June 11 President Kennedy makes a radio and TV address regarding the desegregation of the University of Alabama

June 12 at the Governor George Wallace stands at a door blocking admission of Vivian Malone and James Hood to the University of Alabama.

June 12 Medgar Evers, field representative for the NAACP, is gunned down in his driveway of his Mississippi home.

July 15 Birmingham –Civil Rights protestors are pushed back by fire hoses

August 28 March on Washington for Jobs and Peace

Labor Day CBS debuts expanded 30 minute Evening News with Walter Cronkite

September 9 NBC debuts 30 minute Huntley-Brinkley report.

September 15 A bomb explodes off at the 16th Street Church in Birmingham, Alabama, killing four girls.

October 24 Adlai Stevenson is assaulted following a speech in Dallas.

November 22

- President John F. Kennedy is assassinated by rifle fire while in a motorcade in downtown Dallas. Texas Governor John Connally is seriously wounded.
- Dallas Police Department Officer J.D. Tippit is shot to death when questioning a suspect matching the description of the President's assailant.
- Lee Harvey Oswald is charged with the murders of Officer Tippit and President Kennedy.

November 24 Lee Harvey Oswald is shot to death by Dallas nightclub owner Jack Ruby.

November 25

- President Kennedy is buried at Arlington National Cemetery.
- Officer JD Tippit is buried at Laurel Land Memorial Park, Dallas, Texas.
- Lee Harvey Oswald is buried at Shannon Rose Hill Cemetery, Fort Worth, Texas.

Texas

It was the Friday before Thanksgiving, the last weekend before the holidays, when all the weekends fall like dominoes into the New Year. However, it was not too soon to think of the '64 election season. After securing a nuclear test ban treaty with the Soviet Union, President Kennedy's domestic agenda was stuck in neutral. A tax cut and civil rights legislation met with resistance, especially by southern democrats. After much debate within the White House, there was a conclusion to put off reforms until after the election. In the meantime, the strategy became to build consensus and cooperation, beginning with a road trip.

After a quick visit to Florida, President Kennedy traveled to Texas to make five speeches over two days for what was officially a nonpolitical trip. Unofficially, this was the unannounced 1964 reelection campaign kickoff /fundraising and fence mending expedition. High level democrats of Texas, including Vice President Johnson, Senator Ralph Yarborough and Governor John Connally had been bickering. Each politician was wrestling with their vision and role on the state and national stage. The result was a fractured Texas Democratic Party, threatening to spread nationally, as a crack in windshield glass. The purpose of this trip then, was to smooth over the cracks and, to borrow a phrase from a Broadway show and motion picture of the era, "put on a happy face" for the party platform for 1964.

–And raise a bunch of money. For Texas, Dallas in particular, the attention of a Presidential visit would present an opportunity to show itself as a burgeoning and vital state and shake any lingering prairie stereotypes.

The President would advance his agenda by speaking to government programs on the local level with stops in San Antonio, Houston, Fort Worth, Dallas and Austin. Accompanying the President to Texas for her first political trip since the 1960 campaign was First Lady Jackie Kennedy, in her first public appearances since the death of their infant son, Patrick, in August.

In San Antonio, the President spoke for the dedication of the Aerospace Medical Center at Brooks Air Force Base. In Houston, he spoke of the new national space center and to the League of United Latin American Citizens (LULAC) at the Rice Hotel. First Lady Jackie Kennedy charmed the LULAC attendees by addressing them in Spanish.

The TV President

John F. Kennedy was the first TV President. Presidents Truman Eisenhower had both appeared on television, of course, but JFK took to the medium, as comfortable in front of the camera as FDR had been on radio. His appearance and on-camera presence during the televised debates with Richard Nixon helped him win the election, after all.

It was Pierre Salinger, his press secretaries, who recommended President Kennedy make announcements himself, rather than just issue a statement. As President, JFK had 64 press conferences on LIVE television. His first press conference on TV drew an audience of 65 million viewers. Print reporters came to resent television's encroachment over what had been its domain, reducing the Presidential press conference from substantial to superficial as JFK played to the TV cameras.

Jackie

There had never been a First lady quite like Jacqueline Bouvier Kennedy. After attending some of the finest prep schools, Jacqueline Bouvier attended Vassar, studied at the Sorbonne in Paris, graduating from college in 1951. Her education prepared her for life as wife of a captain of industry, titan of finance, diplomat or, politician. After college, she worked for the Washington Times Herald as "the Inquiring Photographer" and later met and married the junior Senator of Massachusetts, John Kennedy.

Though born into privilege, John F. Kennedy's life was All-American. From his formative years as a teen during the depression, a decorated veteran of World War II and family man at the pinnacle of his field, his life mirrored the American experience. He was only 43 at his inauguration and she, 31. The Kennedys, with their two children came to represent the ideal of the modern American family, at least as far as America pictured itself. John and Jackie Kennedy were the first First Couple in over fifty years who did not look like grandma and grandpa. As often happens, with a new administration, Americans began to reflect the first couple. JFK seldom wore hats, even on the sunniest days, even when golfing or sailing. Sales for men's hats plummeted. Women began dressing like Jackie; she became a favorite for magazine covers, everything from Life, Look, TIME, and Newsweek to celebrity titles.

As First Lady, Jackie Kennedy sought private donations to finance restoration and preservation of the White House. She oversaw all aspects of the project, established a White House committee on Fine Arts, created the post of White House curator and gathered historically significant American art and furniture from around the United States while restoring all the public rooms in the White House. Some pieces were in the possession of former Presidents or their families and others, as near as the White House basement. She wrote and proposed the layout of the souvenir book available at the White House gift shop.

When the redecoration was completed, Charles Collingwood of CBS News and the First Lady served as reporter and guide for a prime time television special program, a tour of the White House. The hour, carried by CBS and NBC (it was broadcast later on ABC) drew an audience of eighty million viewers and earned Mrs. Kennedy an honorary Emmy.

Guests at the White House reflected her passion for the arts. Performances at the White House included the Metropolitan Opera Studio, violinist Isaac Stern, cellist Pablo Casals and modern dancer Martha Graham.

Jackie's appeal was international. She famously upstaged the President when they visited France in May of 1961. The First Lady spoke French so well and so confidently she charmed everywhere she went. In response, the President said at one gathering: "I do not think it altogether inappropriate for me to introduce myself," he said, "I am the man who accompanied Jacqueline Kennedy to Paris."

The President had two beautiful kids and he knew it. Whenever he could get away with it, he'd have the Caroline and John and an official White House photographer. The photos, he knew, would be carried in all the papers. Any photo that tugs on heartstrings stays longer on the mind than a critical story.

Fort Worth, November 22, 1963

The President flew onto Fort Worth, where he spent the night at the Texas Hotel. In the morning was a welcome breakfast by the Fort Worth Chamber of Commerce. His first order of business this rainy morning was to speak to a crowd in front of the Texas Hotel. With potential donors inside for the Chamber of Commerce breakfast, this was a sea of voters, the people who worked 9 to 5. Joining him on the platform built for the occasion were Vice President Lyndon Johnson, Texas Governor John Connally and Senator Ralph Yarborough. Typical of JFK, who would seldom yield to the elements, he was the only one not wearing a raincoat.

Despite a morning chill and light rain, the only disappointment to the assembly of four thousand (some estimates report the crowd was double) was the absence of Jackie. As he stepped to the microphone, he made sure to acknowledge her absence while saluting his audience. "There are no faint hearts in Fort Worth," he began, "and I appreciate your being here this morning. Mrs. Kennedy is organizing herself. It takes longer, but, of course, she looks better than we do when she does it." The crowd responded with laughter and applause.

His speech was about the nation's need for being "second to none" in space exploration and national defense. The President was keen to note the development of two Air Force bombers, the B-58 and the TFX fighter, to be built by General Dynamics, in Fort Worth. After the short speech, the President descended from the platform and, in a characteristic move that drove the Secret Service nuts, went to shake hands with many of the spectators.

Local stations worked together to produce coverage of the day's events. In a cooperative effort, KTVT and WBAP televised the Chamber of Commerce breakfast from the Texas Hotel, making the telecast available to all. Anchoring the live coverage was KTVT news director Ed Herbert. While filling time, Herbert described the route the presidential motorcade would make after breakfast to Carswell Air Force Base for those who wanted a first-hand look. There was nothing unusual about this; the route was in the newspapers.

The President entered the Hall accompanied by "Hail to the Chief", played by the Eastern Hills High School band. Master of ceremonies Raymond Buck introduced organizers and guests such as Vice President Johnson and Governor Connally, each to a round of applause. Buck then announced "And now, I know, is an event you have all been waiting for", which served as the entrance of the First Lady, who entered the ballroom to a standing ovation. As she made her way to the head table, KTVT's Herbert narrated. "As you can tell from the reaction, that's Mrs. Kennedy, she's been standing outside, in the kitchen, waiting for the introduction of the guests at the head table to be completed. Now she is moving up to the head of the ballroom. For the ladies in the audience, she is wearing a pink outfit trimmed in black; if I knew more about it I'd tell you the kind of material, it looks like a nubby material, a wool of some type and with the chilly weather outside today, this probably very appropriate."

Before getting to his prepared text, the President ad-libbed "Two years ago, I introduced myself in Paris by saying I was the man who had accompanied

Mrs. Kennedy to Paris. I'm getting somewhat that same sensation as I travel around Texas." After holding for laughter and applause, he spontaneously added "Nobody wonders what Lyndon and I wear", to genuine laughter and outright guffaws.

The theme of the President's ten minute speech was the defense industry of Fort Worth, through the prism of military preparedness and national security. After noting the defense budget increased by 20% under his administration and how the defense industry in Fort Worth had grown "in the past three years", the President told a little white lie after he listed more quantifications. "I don't recite these for any partisan purpose."

Of course, he did. As with the rest of the Texas speeches, he was telling the crowd not to forget who was President during this rise in productivity and jobs while not telling them he was running for re-election. That was for another time.

Following his remarks, the President was presented with gifts of a cowboy hat and boots. The audience good naturedly gibed him. "Put it on", they said. The President responded with a wave of the hat and pledged "I'll put it on in the White House on Monday, if you come up; you'll have a chance to see it there."

After the breakfast, the presidential party boarded Air Force One to fly to Dallas. It was only a forty minute drive but Air Force One assured protection. Plus, stepping off a plane before an enthusiastic crowd makes for good TV. Vice President Johnson preceded JFK to Dallas via Air Force Two to present the photogenic welcome.

It was Dallas that posed the most challenging. The Stevenson incident was still a fresh wound, reception of the President was unpredictable. Years before the success of the NFL's Cowboys or the glamour of the television series named for the city, Dallas was known as the conservative Capitol of capital, rich in the industries of insurance, banking, transportation, petroleum. Billionaire oilman H.L. Hunt called Dallas home. Evangelical radio preacher Billy James Hargis and his Christian Crusade, the John Birch Society, and National Indignation Committee each had a significant following in Dallas. Dallas was also home to retired U.S. Army Major General Edwin Walker. In conduct unbecoming his office, Walker publicly accused Eleanor Roosevelt and former President Harry Truman of communist sympathies. Walker eventually resigned his commission after attempting to influence votes of those under his command.

On the day the President visited Dallas; the National Indignation Conference
distributed mock "Wanted for Treason" handbills with JFK's photo. An
advertisement with a series of pointed questions directed at the President by
the American Fact Finding Commission carrying the name Bernard Weisman
appeared in the Dallas Morning News.

Adlai

On Thursday, October 24, United States Ambassador to the United Nations Adlai Stevenson addressed an audience of seventeen hundred at the Dallas Municipal Auditorium to mark U.N.Day. His speech was interrupted by hecklers and boos. Exiting the stage, he was mobbed by protesters, at least one of whom spit on him.

One woman, carrying heavy placard, bopped Stevenson on the head with her sign that read "Who Elected You?" She was Cora Frederickson, a 47 year old resident of Dallas. When Stevenson asked her what was wrong she exclaimed ""Why are you like you are? Why don't you understand? If you don't know what's wrong, I don't know why. Everybody else does."

As for striking Stevenson with her placard, Frederickson claimed she "was pushed from behind a negro."

Business leaders sent Stevenson a telegram of remorse. Dallas Mayor Earl Cabell criticized "Right-Wing Fanatics". On Sunday, November 17, a front page story encouraged "An Incident Free day Urged for JFK Visit". Former Vice President Richard Nixon, in Dallas the day before the President's visit, was confident Dallas would provide JFK "a more generous welcome." Stevenson and some members of Congress advised JFK to skip Dallas, but the Dallas-Fort Worth area was too rich in votes and money to ignore.

 Following his stop in Dallas, the President and First Lady would attend a $100 a plate dinner at the Texas Governor's mansion (the only official political event of the trip) and spend the night as the guest of Vice President and Mrs. Johnson at the LBJ ranch near Austin. On Sunday, the President would return to Washington for a Monday meeting with Henry Cabot Lodge Jr., the United States ambassador to South Vietnam.

Television News grows up

Prior to September, 1963 the daily Network newscasts of ABC, CBS and NBC was only fifteen minutes. The staffs of CBS and NBC news had been hounding their programming departments to expand their show to a nightly thirty minute broadcast. The programmers always answered they could never get the affiliates, each network's local stations not owned by the network, to give up the extra quarter hour and commercial revenue that came with it.

On Monday, September 2, Labor Day, a slow news day by any standard, CBS broke through the fifteen minute barrier with a thirty minute edition of The CBS Evening News. The telecast with Walter Cronkite included a special feature, an interview with President John F. Kennedy at his Hyannis Port, Massachusetts, home. NBC's hand was forced and expanded its Huntley-Brinkley Report to a half hour a week later.

JFK was the first President to visit Dallas since Harry Truman's "Whistle Stop" railroad campaign in 1948. For the Dallas trip broadcast networks would carry the story in their nightly newscast but would not go overboard. CBS dispatched reporter Dan Rather from its New Orleans bureau. Otherwise, the news would come over the teletype and the networks would pick up the local coverage. For Texas media, it was "pull out all the stops". For live coverage, local television used a pool arrangement, a reporter and crew of one station to cover for all stations. NBC affiliate WBAP of Fort Worth covered the Chamber of Commerce breakfast. The ABC station, WFAA carried the arrival at Love Field. In addition to televising the arrival of Air Force One, WFAA had eight reporters covering the President's Dallas trip, each serving as their own camera operator for filmed reports they would edit and write for the evening newscast. KRLD was to provide the pool coverage from the Trade Mart luncheon.

Distributing the News

For almost all of the twentieth century, there was in every newsroom an area for the wire service teletype. Introduced in 1914 by the Associated Press for subscribing newspapers, the teletype dispatched news reports transmitted the world over by telephone line. Soon, the United Press and International News Service (later to merge and become United Press International) and other services began their own teletype. The teletype introduced "same day news" to newspapers across the country, such as word of the end of World War I or, the market crash of 1929. A generation later after some squawking by newspapers (fearful of losing readership) and much negotiation, the teletype became a staple of radio news, for the networks such as CBS, NBC, and Mutual Radio and, local stations. Each service had its own teletype, depending on its budget; some newsrooms had three or four teletypes. In newsroom parlance, they were called "the wires."

In the 1930's technology introduced the wire photo, black and white photographs transmitted by sound waves. The wire photo was soon a must-have for every newspaper and a vital boost of the still image as photojournalism. Before the age of satellites and workplace computers, the teletype and wire photo images were a fixture for television newscasts into the 1980's. They were also pretty noisy. A newsroom with several wire machines became a din. On the rare occurrence newsroom management was sympathetic; teletypes had their own room, a converted broom closet or coatroom. –A reporter can always hang his coat on the back of a chair.

As these machines of another time have made way for desktop terminals (not to say anything of tablets, handheld devices and digital transmission), it may be instructive to note wire service machines resemble electric typewriters on steroids, spitting out news 24/7.

Typical of wire service machines of the day, a bell would ring signaling a new news story, an update or, an emergency. A ring of four bells on UPI meant an "Urgent" story. Five bells meant a "Bulletin". For the biggest of stories, ten bells would signal a "FLASH", an emergency, a potential history making occurrence. On November 22, 1963, the wire machines rang ten times.

Dallas, November 22, 1963

People began to arrive as early as 7:30 am to get a glimpse of the President Kennedy's arrival at Dallas' Love Field. By the time of Air Force One touched down four hours later, the sun had broken through, bringing a bright, beautiful sky. Three hundred cops watched of over a crowd of two thousand –and growing- at the airport. When the door to Air Force One opened and the First Lady and President appeared, they saw an enthusiastic public waiting for them.

Covering from Love Field for WFAA was Bob Walker. As the door of the jet opened, Walker exclaimed "There is Mrs. Kennedy and the crowd yells! And the President of the United States and I can see his suntan all the way from here!" he enthused. Walker described Jackie's woolen suit as "nubby", just as Herbert had at breakfast. On black and white television, the suit appeared light but, in person, the shade is reminiscent of the color of strawberry ice cream. (Pamela Turnure, Mrs. Kennedy's Press Secretary made it a point to tell reporters it was NOT pink.) President Kennedy in a dark blue suit followed. They looked every bit as White House Aide Dave Powers said like *"Mr. and Mrs. America"*. As they descended the staircase, the crowd reacted with an explosion of cheers. Many carried the state flag of Texas. One waved the Stars and Bars of the Confederacy. Most signs were welcoming. However, there were a few anti Kennedy, with the message "Lets Barry King John", (a reference to likely 1964 Republican opponent Barry Goldwater) or "Help JFK stamp out Democracy" in the mix. One sign read "Nixon Go Home" and Walker reminded his audience the former Vice President had indeed been in Dallas, but flew out of town that morning.

For the benefit of good pictures for print and TV, the President and First Lady took a moment to shake hands with Vice President and Lady Bird Johnson, Dallas Mayor Earl Cabell and Texas Governor John Connally and Nellie Connally, the Governor's wife. The President surprised the Secret Service and WFAA's Walker, by breaking away from the reception spontaneously to shake hands with airport workers and some of the crowd gathered at the airport's chain link fence. Photographers battled for a good angle as Walker narrated "This is great for the people and makes the eggshells even thinner for the Secret Service, whose job it is to guard the man."

The crowd so congregated the President disappeared from view. Kennedy continued shaking hands took a step or two away from the fence and waved and walked further down the fence, followed by Mrs. Kennedy, to shake more hands. This had gone on for five minutes and yet Walker let the images speak for themselves. "They did not expect this!" he said. As they came closer to the camera, "Now as he becomes more visible down at this end, the people cheer." Each time the Secret Service drew closer, the President would lean into the crowd. "Boy, this is really something!" said Walker. It was seven minutes of this before the Kennedys found their way to the limousine.

When making such public appearances, the Secret Service was most comfortable when the President's limousine had a hard top. For other occasions, the car would be with a clear roof, installed in sections; referred to as the "bubble top". At the President's request, the car was open on such a lovely day. -The better to have everyone see their President and First Lady.

For television to cover the President's motorcade from the airport to the Dallas Trade Mart would stretch the facilities of the TV stations. To mount a TV camera on a truck would pose an encumbrance for the Secret Service and the people who turned out to see the motorcade. Besides, the course from the airport was only eleven miles. Live coverage would resume with the President's arrival at the Trade Mart.

Driving the lede car was Dallas Chief of Police Jesse Curry, followed by the President's car with Secret Service Agent Jim Greer at the wheel. Governor and Mrs. Connally rode with the President and First Lady. Motorcycle officers gave escort. Another vehicle with six Secret Service agents (two on running boards) followed. Then, the Vice President's car (with Senator Yarborough riding along) and the press pool car. The press pool car, courtesy AT&T, featured a rarity for the time, a car phone. Three reporters, from a wire service, radio and TV, served as the media pool on a rotating basis. That afternoon the reporters were Merriman Smith of UPI, Bob Clark from ABC, and Sid Davis of Westinghouse and Jack Bell of the Associated Press. Two buses of the White House press corps took up the rear.

As the motorcade made its way out of Love Field, Walker wrapped up with explaining the motorcades route: to Main, a right turn to Houston Street, a left onto Elm, continuing to the Triple Underpass" (a term that would become familiar to Americans) "and onto the Trade Mart". It was at the Trade Mart telecasts would resume when President Kennedy would speak "At approximately 1o'clock which will also be carried live on most of these channels, and then we'll be back here of course as we told you, approximately at 2:15 for the President's departure."

One television viewer in Dallas was the Very Reverend Oscar Huber. Thinking he would like to see a President in person just once in his life, he turned off the TV and took a walk to enjoy the motorcade.

Bucket Brigade Photojournalism

Most photographers from national publications took the press bus. Some, from local papers or working for the wire services, rode in a motorcade press car. After all, there were afternoon editions to get out. The same was true for reporters shooting moving pictures. When a roll was a done, a photographer would put the spent film in a bag and toss it to Runners situated along the route. The Runners then fulfilled their job description by taking that to the newsroom, lickety- split.

Daytime Television

As their TV stations encouraged their audiences' to view the motorcade firsthand, Dallasites lined the sidewalks. It was estimated over 200,000 people, almost a third of the population, turned out for a glimpse of the President and First Lady.

The rest of the country was served the daily TV fare of game shows, reruns and, daytime dramas. ABC and NBC handed over the daytime schedule to its local stations. Only the CBS network was up, with the popular soap "As the World Turns" for every viewer east of the Rockies.

Breaking the News

The first report of something wrong came over the United Press wire at 12:34 Dallas time, 1:34 in the east. Merriman Smith, the UPI reporter in the White House press pool car behind Kennedy picked up the phone and shouted to an UPI news assistant his bulletin to Wilborn Hampton, a recent college graduate, new on the UPI desk. He sent out the following: Dallas, Nov. 22 (UPI) – THREE SHOTS WERE FIRED AT PRESIDENT KENNEDY'S MOTORCADE TODAY IN DOWNTOWN DALLAS.

At 1:36 eastern, Don Gardiner made the announcement on the ABC radio network. At 1:40, Walter Cronkite read the news on CBS television. Similarly to Cronkite and CBS, announcer Don Pardo broke the news on the NBC from a radio booth for WNBC-TV New York.

As the World Turns

In today's episode, Nancy Hughes is busy knitting a Christmas sweater for her grandson. Her son Bob enters the living room after a long day. After shedding his overcoat, he reports to his mother that he has invited his estranged wife and son for Thanksgiving. Cutting to a scene set the following morning, Nancy is dusting the bookshelf and reports to her father that he can expect his great grandson for the holiday dinner.

An otherwise routine episode of an otherwise routine soap opera was interrupted ten minutes in with a CBS News Bulletin slide.

Walter Cronkite's first announcement from a radio booth patched to the television network. In 1963, the New York headquarters for the CBS Evening News was above Grand Central Terminal. A long, enclosed catwalk separated the studio from the newsroom. It would take some time to move a camera into the newsroom, warm it up and, get on the air.

"Here is a bulletin from CBS News," Cronkite begins. "In Dallas, Texas, three shots were fired at President Kennedy's motorcade in downtown Dallas. The first reports say that President Kennedy has been seriously wounded by this shooting." There is a pause and it is soon apparent someone is handing Cronkite more pages. "More details just arrived. These details as previously, President Kennedy shot today just as his motorcade left downtown Dallas. Mrs. Kennedy jumped up and grabbed Mr. Kennedy, she called 'Oh, no!' the motorcade sped on. United Press says that the wounds for President Kennedy perhaps could be fatal. Repeating, a bulletin from CBS News, President Kennedy has been shot by a would-be assassin in Dallas, Texas. Stay tuned to CBS for further details."

As Cronkite was making this announcement, CBS technicians are moving a heavy, bulky studio camera to the newsroom, outfitting Cronkite's desk with a microphone and setting up a light kit.

Onscreen, CBS cuts from the bulletin to a commercial break for Nescafe instant coffee. Then there is a station identification and promotion for that evening's episode of the drama "Route 66" As the show was about to resume, back to the CBS Bulletin slide and the voice of Walter Cronkite.

"Here is a bulletin from CBS News. Further details on an assassination attempt against President Kennedy in Dallas, Texas. President Kennedy was shot as he drove from Dallas airport to downtown Dallas. Governor Connally of Texas, in the car with him was also shot. It is reported three bullets rang out. A secret service man has been…was heard to shout from the car "he's dead." Whether he referred to President Kennedy or not is not yet known. The President, cradled in the arms of his wife, Mrs. Kennedy, was carried to an ambulance and the car rushed to Parkland hospital outside Dallas. The President was taken to an emergency room in the hospital. Other White House officials were in doubt… in the corridors in the hospital as to the condition of President Kennedy. Repeating this bulletin: President Kennedy shot while driving in an open car from the airport in Dallas, Texas to downtown Dallas. Recounting again the details of this incident, three shots were heard to ring out as Kennedy and Governor Connally and Mrs. Kennedy rode in the backseat of the open car. Immediately, a Secret Service man said he saw blood spurt from the President's head. He fell into the laps (sic) of Mrs. Kennedy and Mrs. Kennedy shouted 'Oh, no'. Governor Connally was seen to crumple also. The car sped on; the motorcade speeded up, rushed to Parkland hospital in Dallas where the President and Governor Connally were rushed to the emergency room. A Secret Service man was heard to say 'he's dead' as the President was lifted from the rear of the White House touring car, the famous bubble top from Washington, and taken into the hospital. Reporters about five lengths behind the Chief Executive heard what sounded like three burst of gunfire. We will keep you advised as more details come in; the incident has taken place only in the last few minutes in Dallas. Stay tuned to CBS News for further details. "

The network entertainment side reverts as if there has been no interruption and As the World Turns resumes as Bob Hughes is now at a restaurant meeting with Dr. Cassen, their conversation punctuated with dramatic pause and cigarette smoke. As the waiter brings menus, the scene fades for another commercial break. Soon back-to-back ads for Friskies dog food are interrupted for a third bulletin. It is now 1:47 eastern time. Over the bulletin slide, Cronkite announces "Here is a bulletin from CBS News: President Kennedy has been the victim of assassin's bullets in Dallas, Texas. It is not known as yet whether the President survived the attack against him. The incident was this: the President, Mrs. Kennedy and Governor Connally of Texas, driving in the famed bubble-top car, from Dallas airport to downtown Dallas, where the President was to make a speech. Three bullet shots were heard to ring out, the President slumped into the lap of Mrs. Kennedy, witnesses say they saw blood streaming from his head. Governor Connally slumped into the bottom of the car, bullet wounds were seen to be in his chest. The car itself rushed on as Mrs. Kennedy was heard to say "Oh, no." It took him to a nearby hospital, the Parkland hospital where the President and Governor Connally were taken into the emergency room, and witnesses there refused to comment whether the President was still alive or not. As the bullets were heard to wing out of the assassin's gun, Secret Service men unlimbered their rifles but, the damage of course by then, had been done."

"As the details come in, let us read a few to you, from the press service wires, reaching us only now, from Dallas. After these three loud bursts of gunshot, Dallas motorcycle officers escorting the President quickly leaped from their bikes and raced up a grassy hill. At the top of the hill, a man and woman appeared, huddled on the ground. In the turmoil, it has been impossible so far to determine whether the Secret Service and Dallas police returned the gunfire that struck down Kennedy and Connally or whether this couple at the top of the hill, crouched down, their inert forms, uh, as the police rushed them, were the would be assassins. Stay tuned to for further details from CBS News."

There is a pause and then, noise of the booth's door opening and paper exchanging hands. Cronkite continues to read from newswire reports, repeat details and announce Congressman Albert Thomas of Houston reports the President and Governor Connally are alive, at least as of a few minutes ago.

He then ad-libs attempts on the lives of Presidents Truman and Roosevelt. Cronkite then announces the CBS network will take a ten second pause (after which he appears on camera). It's now 2 p.m. eastern.

Unbeknownst to the home viewer, the motorcade proceeded to the Trade Mart. It was only there that most of the press corps heard the news and headed for the hospital.

Fearing the worst, Reverend Huber makes his way to his car. Fighting traffic, he and a colleague drive to Parkland Hospital to offer what help they can.

Dan Rather, the CBS News reporter in town for the story, ran back to KRLD. Program director Jay Watson of WFAA, taking in the motorcade off-duty, raced from his vantage point to the studio. Watson was on the air within minutes, interrupting WFAA's Julie Bennell Variety Show to break the story on television in Texas.

Switching Channels for Live, Breaking News

NBC

As the news of the assassination attempt came over the newswires, NBC was in local programming. In the east, NBC stations were airing a rerun of the John Forsythe sitcom; "Bachelor Father". The network newsroom was no more active than any other Friday afternoon, with some just back from lunch. When a newsroom hand at NBC's headquarters at 30 Rockefeller Center heard the ten bells, he called out for a reporter. Across the newsroom at his desk correspondent, Bill Ryan was writing copy for the hourly network radio newscast. He stood up from his desk and shouted "What do you need?" "Get back to TV right away!" the news hand said. "The president has been shot!"

Ryan headed to Studio 5HN, the NBC "Flash" studio. A small, bare facility with cheap wood panel walls and a studio camera, it exists for just such a day. Chet Huntley, half of the network's "Huntley-Brinkley Report" and Frank McGee joined Ryan. It took until 1:53 for NBC to clear its local stations and get the network on the air. Huntley, McGee and Ryan begin to report. It is not until 1:57 that someone starts a videotape recording.

David Brinkley saw the news cross the wire in NBC in Washington and looked for management. For the moment at least, no one was around.

CBS

The CBS TV network returns at 2 p.m. with a station identification and its CBS eye-in-the-clouds logo. After the CBS News Bulletin slide, Cronkite appears at his desk. He is wearing a tie and white shirt, but not the jacket to his suit. Cronkite has the first word from the Parkland Emergency Department; Congressman Jim Wright has said Kennedy and Connally are still alive. Cronkite tosses to Eddie Barker, CBS affiliate KRLD news director who doubles as a reporter from the Dallas Trade Mart, Kennedy's lunchtime speech destination. Two cameras provide coverage, one from a balcony, and another from the floor.

An unconfirmed report has Kennedy critical, Connally entering surgery and a Secret Service agent dead. Barker reports "we now hear it is a man and a woman" are the suspected assassins. CBS switches back to Cronkite who is handed a wire photo of the President and Jackie Kennedy in the motorcade moments before the shooting. As Cronkite explains the photo, producer Don Hewitt gives him the cue the network will switch back to Dallas.

The camera is fixed upon the Reverend Luther Holcomb, Executive Secretary of the Dallas Council of Churches as he concludes a prayer for the President then pans right, giving a sweeping view of the hall. Barker reports "As you can imagine, there are many stories coming in now, as to the actual condition of the President. One is that he is dead. This cannot be confirmed. Another is that Governor Connally is in the operating room. This we have not confirmed." As the camera zooms in on an African-American waiter, standing before the Seal of the President where he would have given his speech, wiping tears from his face, Barker continues "Another is, and apparently, this is correct that one of the Secret Service agents, whose job it was to guard the life of the President, was killed in his line of duty." As attendees begin filing out of the Trade Mart, Barker explains the Trade Mart is roughly three miles from where the President was shot; Parkland hospital is about a mile and a half from the site of the shooting. He also stresses the reliance of scanty reports this early in a developing story is an ephemeral proposition.

Switching back to New York, Cronkite reports blood transfusions are being given to President Kennedy and that two priests are at the emergency room, as are many members of the White House staff. With the report of an arrest, Cronkite leads to a KRLD audio tape report of a witness who reports he saw a "colored" man shoot at the motorcade. At the end of the radio report, Cronkite reports the New York Stock Exchange has closed. Off camera, producer Don Hewitt informs Cronkite "KRLD is reporting they've been told from the hospital the President is dead, only a rumor." Looking into the camera, Cronkite reminds the viewer "Well, that is something you've been told a moment ago directly from KRLD television in Dallas, and that is the rumor that has reached them at the hotel, that the President is dead. Totally unconfirmed apparently, as yet."

Switching back to the Trade Mart, Barker again reports an unconfirmed report the President is dead and interrupts himself "Word just came to us, a minute ago. The word we have is that President Kennedy is dead. This we do not know for a fact" but Barker reports he heard this from a doctor at the hospital. "This is still not officially confirmed but as I say, the source would normally be a good one." Guests are milling about and technicians are seen removing microphone wires and the Presidential seal from the lectern.

Reverend Holcomb again steps to the microphone the same technicians reattach cables. What was a matter of expediency, the breakdown the lectern and microphones upsets the audio levels for television and for the moment, neither Reverend Holcomb nor KRLD's Barker sounds clear. Pastor Fletcher Dickinson offers a prayer to the after which Barker repeats "The word was not given to the crowd but we understand that the President is dead. This, we do not have an absolute confirmation on but, this is the word that we have."

Cronkite adds that Vice President Lyndon Johnson, in the car immediately following the Presidential limousine, was seen holding his arm as he entered the hospital. He has not been seen in the hospital corridors, perhaps he has wounds too? Handed a note, he reports that CBS' Dan Rather in Dallas reports the President is dead, but Cronkite resists committing until official word from the White House.

Both speaking extemporaneously and relying on wire service copy; Cronkite removes his glasses when looking into the camera and relies on the specs when he reads. With new information handed to him, Cronkite reports reporters at the hospital say Mrs. Johnson says Vice President Johnson has not been hurt. "We have no further information on the condition of Governor John Connally of Texas, who was shot in the chest by this assassin's bullets; he was taken to another room at the hospital and no further information on him so far." Cronkite is handed another sheet of paper with a report from Bill

Stinson of Governor Connally's staff. Stinson is quoted Governor Connally was shot "just below the shoulder blade, in the back." When asked what happened, Stinson quoted Connally as saying "I don't know, I guess from the back they got the President, too." Cronkite then says "Governor Connally could very well have possibly have been shot in the back with the assassin's bullet coming from the front of the car."

Two minutes later, a young man peels a story from a service wire teletype and quickly drops it on Conkrite's desk. When he picks it up, he reads "The two priests who were with Kennedy say he is dead of his bullet wounds." With his glasses off, he adds "That seems to be as close to official as we can get at this time."

ABC

Of the three broadcast networks of 1963, ABC lagged behind its older competitors. While ABC's prime time lineup was competitive, the ABC network had had fewer affiliated stations and the network news division had neither the manpower nor budget of CBS and NBC. While those two networks expanded their evening newscasts in September, ABC stayed with a fifteen minute format until 1965.

Anchorman Ron Cochran was summoned from a restaurant lunch when the story broke. As he reads wire service reports, stagehands struggle to provide a backdrop as he stands before a bare wall. Ed Silverman joins Cochran on the set to relay news by telephone from reporter Bill Lord, covering the story from the Sherriff's department opposite the scene of the shooting. After each update, Lord would need to put the phone down and get more information. A switch for comment from ABC's Edward P. Morgan, reporting from Washington is frustrated by an audio glitch. ABC has to dump the report. When Morgan finally reports from the newsroom, the clicking and clacking of the wire service machines can be heard.

NBC

McGee is holding a telephone in anticipation of a call from Robert MacNeil, who was on the press bus. After hearing the shots, MacNeil left the press bus and went to the nearest building, the Texas School Book Depository, in search of a pay phone. After his first report, MacNeil made his way to Parkland Hospital. There, reporters gather in a hospital classroom or roam the halls for a pay phone. In what must have been a television first, McGee accepts a collect call from NBC's Robert MacNeil on live national television. When it is apparent there is no connection to broadcast MacNeil's voice, McGee has to repeat MacNeil, a sentence at a time. Staffers can be heard whispering instructions as someone hands a device to Huntley, who puts it in front of McGee. Ryan says "Put the receiver on that Frank and apparently, the receiver will act as a speaker."

The result is nasty, piercing, feedback. For a minute or two, the network is halted by want of a functioning audio patch. MacNeil gets a med student to mind the pay phone as he gets more information and the call mercilessly concludes. Ryan announces the news of JFK's receiving blood transfusions as a production assistant holds up a wire photo of the President and First Lady in the motorcade, hastily pasted onto cardboard.

With wire service information that underlines just how disjointed and incomprehensive the news from Dallas was, Huntley reports "We have this information which adds up to something rather inconclusive. It says the Secret Service says that the President remained in the emergency room and the governor was moved to the general operating room of Parkland Hospital. One Secret Service man was overheard telling another that there was no need to move the President because emergency facilities were entirely adequate in the emergency room." Huntley stops reading and looks into the camera "That would indicate, just as a sort of snap judgment evaluation, that Governor Connally was worse wounded than the President but as I say, that is only a snap judgment evaluation." Ryan adds "Or it could mean, Chet, that they are separating the two cases so that each can be given full attention."

With that, Ryan recaps and tosses to reporter Charles Murphy, at NBC station WBAP, Fort Worth.

As was ABC and CBS, NBC's coverage was in black and white, still the standard. So when Murphy appears in color, it is a bit of a surprise. (WBAP was the first station in Texas to broadcast in color.) Murphy reports a young man has been taken into custody. Before he can report further the audio drops out, muting Murphy. NBC dumps the feed, fleetingly returns to Ryan, back to Murphy. This time, the only sound is ear splitting reverberation. The shot returns to Ryan who explains "As you can appreciate, communication facilities, as you just saw, went in and out. This is a time of what can best be described as controlled panic. The arrangement of that switch to Fort Worth were made entirely hastily, under conditions of extreme pressure and that is why the picture came and the audio didn't and when the picture dropped, the audio came in. "

While the connection to WBAP is sorted out, Ryan recaps. "The President and Governor Connally were shot in an assassination attempt while riding in a motorcade through downtown Dallas, First Lady Jacqueline Kennedy was not hurt. There is no indication of where the President and Governor were hit or how often. There were reports of a man spotted in a window with a high powered rifle but he has not been found." He then leads to Charles Murphy again.

The third time is the charm as Murphy reports a man in his early twenties has been taken into custody, to the County Sherriff's office, which happens to be near where the shooting took place. The individual vehemently denies involvement. Two Roman Catholic Priests have been called to the Presidents side, the President has received blood transfusions and the Governor has been moved to an operating room. Murphy tosses back to New York with "Charles Murphy reporting, from WBAP, Fort Worth/Dallas."

Ryan reports of a call from Parkland Hospital for B+ blood and then tosses to David Brinkley, reporting from NBC in Washington, D.C. With an economy of words that marked his career, Brinkley reported briefly, ending with "There is no particular or unexpected reaction here yet, because most of Washington has just found out and is shocked and stunned along with everyone else. The White House has no more information at the moment than we have. We'll stay here of course; ready to bring you more the minute it arrives. In the meantime, we'll switch back to you. Chet, Frank, Bill."

Brinkley said each name because he did not know which of the three would take his handoff. It was also his cue to the Washington control room to go to New York as the switch was made manually.

When another attempt to switch to WBAP fails, Ryan ad libs a recap and eventually tosses to Tom Whelan at WBAP. Whelan has an audio interview with Jean Hill, an eyewitness to the shooting. Mrs. Hill reports that she heard the shots as she was looking at the President, that she thought the shots were fired from a hill near the triple underpass an area to become known as the "Grassy Knoll." In addition to having someone in custody, Whelan reports "it is believed" police have recovered a rifle from the Book Depository.

Without having to say, each network acknowledged a decision made independently and simultaneously: each network was staying with the story, eschewing entertainment programs and commercial announcements.

Communications context

Robert MacNeil has returned to the payphone but, NBC still has not worked out the bugs. Ryan gets the cue to tell McGee the phone link is fine but it is soon painfully apparent it is not. Rather than try the attachment again, McGee advises MacNeil he will once again serve as relay. The camera stays on a two shot, McGee and Huntley. Huntley is holding an earpiece so that he may hear MacNeil.

Reading from wire service copy, Ryan reports the known details of Connally's wound below the shoulder blade. Huntley refers to JFK's breakfast remarks. This would be an opportunity for a producer to run thirty seconds of video of the President at the breakfast. After all, so far NBC had three men talking on set, one of them on the phone and some photos on cardboard. But in the days instant live shots, feeding video from Dallas would have taken several people off the tasks at hand. Plus, all phone lines, the communications grid of the day, were tied up. All news institutions were calling bureau to bureau. Federal and state law enforcement was also on the phone. The FBI was calling Dallas; the CIA was calling the FBI, the National Security Agency was calling the CIA, Dallas police were calling everybody and, in Washington, federal employees from janitors to elected officials were calling to say they would not be home anytime soon. Across the country, neighbor was calling neighbor to ask "Have you heard?" Phone traffic in DC did not ease until all non-essential government employees were told to go home.

It also bears mention, each network anchor, Cochran on ABC, Cronkite on CBS and, Huntley, McGee and Ryan on NBC, were on the air without the now ubiquitous anchorman's earpiece, the IFB, (Interruptible Feedback) the link between news producer and reporter.

Now juggling two phones, the camera pulls back to McGee, who reports neurosurgeons have been called in "indicating one of the two had been hit in the head," "Either the head or possibly some spinal damage," Ryan adds. "There is also this,' Ryan says. The camera pans right and pushes in to Ryan, reporting from wire service copy, "Understandably, in a situation like this, the information comes in fragments and comes from unexpected places and from uncontrolled angles. Senator Ralph Yarborough, a Texas Democrat who was riding in a nearby car when the attack took place on the President, said he saw the President's lips moving, at what he called 'a normal rate of speed' while Mr. Kennedy was being rushed to the hospital. How much it means, we do not know."

The first, unspoken law of television news is to keep advancing the story. On a crowded, three anchormen set with television's first story of crisis, the challenge of advancing the story is to report information as it comes in and then, pass the story on. Much like a volleyball during a game, the story must remain in motion; each reporter must be allowed to move it along and get set for the next time it comes their way. That is what is happening here.

McGee adds, "There is further word from the hospital, Bill that they are trying to make arrangements as quickly as they can for a press conference where as much actual and detailed information as they have can be disseminated." "And there is word here Frank" Ryan replies, "at least one neurosurgeon has arrived at the hospital. I should imagine in a case such as this that virtually every medical specialist of any sort and description and capability would be called in to the hospital so that all medical treatment would be available to the President. Chet?"

Huntley introduces another angle when he says "I would assume that every person listening has flashed back to that day in April, 1945 when Franklin Delano Roosevelt…" "Excuse me Chet," Ryan interrupts. "Here is a Flash from the Associated Press, dateline Dallas. Two priests who were with President Kennedy say he is dead of bullet wounds. There is no further confirmation, but this is what we have on a Flash basis from the Associated Press. Two priests in Dallas who were with President Kennedy say he is dead of bullet wounds. There is no further confirmation; this is the only word we have indicating the President may in fact have lost his life. It has just moved on the Associated Press wires from Dallas. The two Priests were called to the hospital to administer the last rights of the Roman Catholic Church and, it is from them we get the word that the President has died; that the bullet wounds inflicted on him as he rode in a motorcade through downtown Dallas have been fatal. We would remind you there is no official confirmation of this from any source as yet."

McGee reports that the Vice President and Mrs. Johnson, under Secret Service protection, have left the hospital. Huntley adds "We must stand by for confirmation, as Bill has emphasized, this is rather sketchy information. We will stand by, we should have…" Ryan interrupts, "Now, uh, there apparently is word that this AP Flash, this report from the two priests that the

President has died of bullet wounds is confirmed, we will attempt to get to station WBAP in Fort Worth/Dallas."

From WBAP comes the report the Dallas police department has informed its officers the President is dead, the Associated Press Flash moved approximately three minutes later. He concludes with "Charles Murphy, returning now, to NBC in New York."

As the network returns, McGee announces he has MacNeil in Dallas. Again, McGee has to repeat his MacNeil colleague and then, the audio patch works. McGee reports that MacNeil reports "White House Press Secretary Malcolm Kilduff has just announced President Kennedy died about 1 o'clock Central Standard time." McGee adds "Which was about 35 minutes ago." It is then when NBC finds MacNeil's voice. "The President died approximately twenty five minutes after the attack took place." MacNeil also reports Vice President Johnson was taken by police escort from the hospital "to assume the Constitutional responsibilities of the Presidency." He also reports a casket has been brought for the slain President. He contrasts this news by noting the Dallas visit produced the largest, most enthusiastic turnout of the President's trip. "The shooting came, according to eyewitnesses, from the second floor of a building called the Texas school book depository, which is about 100 yards from the tree-lined parkway on which the President was driving when the shooting returned. Three shots were heard."

There is news a Dallas police officer has been shot and killed.

CBS

The most enduring clip of this day is from CBS and Walter Cronkite. It is without technical glitches. It is also elegant. Cronkite is handed a piece of paper. "From Dallas, Texas, the Flash, apparently official, President John Kennedy died at 1pm, Central Standard time, 2 o'clock Eastern Standard time, some 38 minutes ago. Vice President Johnson has left the hospital in Dallas. We do not know to where he has proceeded. Presumably, he will be taking the oath of office shortly and become the 36th President of the United States."

While ABC and NBC relied on two and three men on camera on CBS, the burden was all on Walter. About ten minutes later, Cronkite is relieved at the CBS desk by Charles Collingwood it is only after Cronkite gets up he realizes the jacket to his suit's has been on the back of his chair the whole time. Bob Trout joins Collingwood and reports his observation s of walking about Madison and Fifth Avenues. Passing cars have radios on. At traffic lights, pedestrians gather. To hear the car radios. A nearby school let out, kids are crying. Trout leaves, he is replaced by Eric Sevareid, who notes the assassins of Lincoln, Garfield and McKinley acted on their own, not acting on behalf of any political entity.

 On NBC, Huntley makes his exit when the camera is tight on Ryan. Both Cronkite and Huntley wait until they are off camera to let their guard down and then collect themselves. Each has their nightly newscast to prepare for.

Collingwood has news a suspect has been arrested and Switch to Neil Strawser in Washington, who speaks of the mood in Washington and presidential succession. There is video with no sound from the United Nations of Adlai Stevenson receiving condolences on the floor of General Assembly.

Cronkite resumes the anchor spot after thirty minutes, now wearing his suit coat. CBS switches to Dan Rather in Texas, who discusses the shock of the news. Back to New York and Cronkite introduces Bernard Eismann, conducting Man on the Street interviews in NYC. Then back to Rather who narrates a film of President Kennedy in Dallas. Sevareid joins Cronkite at his desk for a quick biography of Lyndon Johnson. Russ Bensley conducts man on the street interviews from Chicago. Dan Rather has the first film of Lee Harvey Oswald in custody. CBS runs a filmed reaction from former President Eisenhower. Switching back to Dallas, Rather interviews CBS White House correspondent Robert Pierpont, who had been on the Press Bus. Going by his notes and observation, he describes first bullet as fatal to the President, to the back of the head and out the throat. There is a live picture of services of New York's St. Patrick's Cathedral. CBS has Adlai Stevenson on film. "The Vice President..." he catches himself "the habit is still strong with all of us. The new President..."

ABC

What ABC lacked in alacrity is offset by unique perspective. Joining Ron Cochran on the set is ABC Vice President for News and Special Events, Jim Hagerty. As the press secretary for President Eisenhower's two terms, Hagerty offers a methodical explanation of what precautions are made for a Presidential trip.

NBC

After Chet Huntley recaps the story, NBC goes to Charles Murphy of WBAP who reports police officer JD Tippit has been killed by gunshot just two miles from the President's assassination.

Until it is determined where Lyndon Johnson is and when or where he will be sworn in, the networks are left to offer recaps and reactions.

The Suspect

The reports of an arrest of a suspect in the shooting death of Officer Tippit are followed with his amazing life story.

Lee Harvey Oswald, a high school dropout, enlisted in the Marines at age 17. While serving in Japan, Oswald was granted a family hardship discharge, to tend to his ailing Mother. Instead of returning stateside, Oswald immigrated to the Soviet Union and announced his desire to become a citizen, resulting in a dishonorable discharge from the Marine reserves. He married a local woman; they had a daughter. Disenchanted with Soviet living, Oswald in 1962 cabled the State Department for a loan to return to the United States. After spending the summer of 1963 in New Orleans as one man operator of the "Fair Play for Cuba Committee ", Oswald settled in Dallas with his wife and (now two) daughters. In October, he was hired as a clerk at the Texas School Book Depository.

Tracing Oswald's steps that Friday have led to the conclusion that, after he murdered President Kennedy, Oswald slipped out of the Book Depository and hopped on the nearest bus. When the bus was held up in traffic, Oswald stepped off and hailed a cab. Getting out before the cab reached his destination, Oswald briskly walked into his rooming house, where he picked up a .38 revolver and a light jacket. –His landlady later told police he left as quickly as he entered, without a word.

In the meantime, a description of the suspect in the shooting of the President went over police radio. Spotting a man matching the description, Patrolman J.D. Tippit stopped Oswald for a few questions. Unsatisfied, Tippit got out of his cruiser and stepped around the front of the vehicle to ask more questions. Oswald pulled out his gun and shot Officer Tippit three times, killing him.

Still on foot, Oswald ditched his jacket and wandered from his residential neighborhood to business district. When another police car neared, he hid in a shoe store alcove before sneaking into the movie house next door, the Texas Theater. The store clerk followed Oswald in as he told the box office cashier to call the cops.

When the theater lights were turned on, Oswald stood and aimed a revolver at an officer. He never got off a shot. WFAA-TV scored an exclusive when Ron Reiland filmed police hauling Oswald away, kicking and screaming. It was not long after his arrest that Dallas Police would describe Oswald as the "prime suspect" of Kennedy's assassination.

CBS switches for a live shot from Los Angeles, a man on the street interview with reporter Charles Kuralt at Hollywood and Vine. Glitches victimize CBS; there is a burst or two of Reverb for a moment. Once the Kuralt interviews conclude, Collingwood reads reactions of Winston Churchill and French President de Gaulle. Neill Strawser in Washington has the reaction from congressional leaders on film and George Herman reports live from the White House with an outline of the next 24 hours.

And so it continues for ABC and NBC.

TV in '63

Videotaped coverage was new; videotape cameras are not portable. Most TV reporters in Dallas carried 16 mm film cameras that record without sound. Reporters working with a film camera that record sound can capture more but require two men. In each case, the film must be brought to the bureau and developed and edited for presentation. To present a filmed report with a narrative track takes more time. Once film is processed, the reporter who shot the film would narrate on the set.

The networks and its local stations attempt to produce live telecasts from Parkland Hospital, the Dallas county jail and, the White House. For Parkland and the jail, it will take time. Cameras for live telecasts are big and unwieldy and require long, heavy cables and rely on connections to lines run by the phone company.

NBC

NBC's Ryan leads to Martin Agronsky who reports flags at the capitol are at half-staff. At the United Nations; Secretary General U Thant offers condolences and a minute of silence.

WBAP's Tom Whalen reports the discovery of a rifle believed to be the murder weapon at the Book Depository. McGee and Ryan confirm the shooting of a Dallas police officer and the erroneous report of a Secret Service agent is dead. A suspect in the shooting of the officer is now in custody. In Washington, Brinkley narrates a live shot of mourners gathered at Lafayette Park, across Pennsylvania Avenue from the White House. WNBC's Jeff Pond interviews New Yorkers near the Rockefeller Plaza. Via phone from Rome, Irving R. Levine reports of reaction of Pope Paul VI. After switching to and from Washington, Ryan announces Lyndon Johnson has been sworn in as President and is en route to Washington. There are more phone reports from NBC's foreign bureaus, Leif Eid in Toronto, John Chancellor from West Berlin and Welles Hangen from Bonn, West Germany. WBAP airs film recorded during the motorcade. Among the cheering throng, a young man is holding a sign that reads "Yankee, Go Home".

The White House announces Lyndon Johnson has taken the oath of office to become the 36th President of the United States aboard Air Force One, just before departure from Dallas. White House photographer Cecil Stoughton captures the moment. Stoughton gives the film to Westinghouse correspondent Sid Davis, who makes sure to make it to the wire services. Acting as the pool reporter, Davis briefs his fellow reporters the Love Field terminal.

Each network reports Secretary of State Dean Rusk and other members of President Kennedy's cabinet, en route to Japan for a conference, had the plane make a midcourse change to return to Hawaii for refueling then, on to Washington.

The first detail of remembrance ceremonies is announced by the White House. Visitation for President Kennedy would be in the East Room of the White House. Saturday would be a day for family, friends and members of the various branches of government. The casket will be moved to the Capitol Rotunda for public visitation Sunday.

ABC

On ABC, Bob Young has taken over for Ron Cochran. He tosses to Bob Fleming of ABC Washington, speaking extemporaneously of young LBJ before switching to Edward P. Morgan outside the White House, who ad-libs on JFK as Senator, Democratic nominee in 1960 and, President. Then ABC runs a taped, sit down interview with network reporters and LBJ, a re-airing of a program from the previous March. Viewers got a look at calm, almost somnolent Johnson taking questions from Bill Lawrence, John Rofson and Edward P. Morgan.

At WFAA, Jay Watson scored a coup, not once but twice. Watson interviewed witnesses Bill & Gayle Newman (with their two young children) 15 minutes after the President was shot. Watson also interviewed dressmaker Abraham Zapruder who, using his new home movie camera, made the only recording of the assassination. At first, Watson told viewers the film was being processed for telecast. Watson spoke too soon. Zapruder's film was 8MM and color; a format not compatible for TV. Eventually, Zapruder's film was developed at a Dallas Kodak lab. After screening the film for the Secret Service and some reporters, Zapruder sold all rights to his film to Life, which ran photos from the movie subsequent issues.

In New York, Howard K. Smith takes to the air for ABC. Smith was putting in a long day. He was fresh off a transatlantic flight, returning from an interview with Egypt's President Nasser. Smith's time on the studio set was brief. Exhausted, he flew home to Washington, got a night's rest and was on the air the next morning.

Friday night

On a WFAA panel discussion, Bill Lord estimates the he saw twenty reporters at Dallas Police headquarters at 2 p.m., ten times as many by 4 p.m. There would be more.

As the story broke, radio and television news departments, newspapers and magazines dispatched reporters from all over the country. Life magazine sends a contingent from its West Coast office. NBC correspondents Tom Pettit, Sander Vanocur, an editor and producer and two camera crews were sent from KNBC in Burbank. (Vanocur would fly on to Washington; he was NBC's main White House reporter.) Routine safety procedures were ignored for a flight from Los Angeles to Dallas. So many reporters with so much equipment were on board, cameras and other broadcasting gear was stowed in the aisle.

A flight of reporters from New York to Dallas made an unscheduled stop in Washington. There was no announcement, but many who boarded in DC were FBI. Travelling to Dallas presented a problem: as long as they were en route, they were cut off from the story. Pilots would offer some news relayed by ground controllers listening to the radio but, by then the news was at least third hand.

NBC in Washington assigns Nancy Dickerson to report live coverage of the return of Air Force One to Andrews Air Force Base. The only person with an available vehicle to drive her does not have a car radio; she is as cut off as her colleagues in the air. The lack of information makes her ride frustratingly quiet.

Security at Andrews is tight; Dickerson gains entry when she shows identification. She finds she is there before any other TV reporters. A WRC-TV truck with one camera for live broadcasts happens to be on the Beltway, the circuit of I-495 that surrounds the Capitol. By two-way radio, it is directed to Andrews. At the gate, Dickerson vouches for her colleagues. As the time for the jet to land draws near, Security is not allowing entrance to other TV trucks; making WRC the video feed for the networks. Dickerson and Robert Anthony will report, WRC's Max Schindler will direct from the TV truck. CBS carried the pictures from NBC, with Harry Reasoner from the anchor desk in New York and Charles Von Fremd reporting from Andrews.

As Air Force Colonel James Swindal prepares to land Air Force One, he radios the control tower. The lights for TV are blinding him; he cannot land until they are turned off. The jet breaks through the darkness. The first thing viewers hear is a production assistant calling to Abernathy "We understand the President changed his mind Bob, the President is not going to talk. Abernathy begins "Several thousand people have turned out, waiting for the President to arrive at Andrews Air Force base." Dickerson picks up "Members of Congress have arrived, leaders of both the House and Senate, their wives are here, and the wives are in tears." The TV lights come on as the jet taxis. Dickerson reports an honor guard from Bolling Air Force Base is present, to meet the fallen President. As the camera stays tight on Air Force One, she continues "An Army helicopter has just landed; it's making a great deal of noise."

When Air Force One comes to a stop, Attorney General Robert Kennedy bounds up the staircase as soon as it is wheeled to the door and disappears inside the plane. A moment later, a lift moves to the rear of the plane and staffers begin moving the casket. The lift descends; soon it is surrounded by men in uniform. The casket is slowly, methodically transferred to a U.S.

Navy ambulance, to be taken to Bethesda Naval Hospital. Two thousand people congregated at Andrews. This is the first public appearance of Jackie and to see bloodstains on her lovely suit is shocking. Next to her is her brother in law, the Attorney General. After Jackie and Bobby get in the ambulance, it drives off. Abernathy and Dickerson describe the scene. The lift truck moves away, replaced with airplane stairs. On the tarmac are many vehicles and military personnel. Some step with purpose, others are not sure what to do. Neither is director in the WRC truck, Max Schindler. He is unaware President Johnson was on the same flight. As the ambulance leaves for Bethesda Naval Hospital, Schindler directs his camera to find a spot and stay on it.

The camera focuses on the rear of Air Force One. President Lyndon Johnson and the new First Lady descend the staircase, followed by their staff. The new President and n first lady receive greetings and he walks to a well-lighted spot with ten microphone stands.

"This is a sad time for all people. We have suffered a loss that cannot be weighed. For me, it is a deep personal tragedy. I know that the world shares the sorrow that Mrs. Kennedy and her family bear. I will do my best. That is all I can do. I ask for your help--and God's."

President Johnson and his wife then take the Army helicopter to the White House. Reporters announce the new President will go to the White House where he will meet with congressional leaders, members of the national security team, the Central Intelligence Agency chief John McCone and, when they return, the Cabinet.

Once the military helicopter ferrying LBJ to the White House is airborne, CBS cuts to Harry Reasoner at the news desk. Reacting to President Johnson's remarks, Reasoner observed "You have seen, live from Andrews Air Force Base in Maryland, the arrival of the plane from Dallas, Texas, carrying the body of the late President Kennedy and also bringing Mrs. Kennedy and new President, Lyndon Baines Johnson. The advances in technology which permit historic events like this to be seen by people as they happen also bring some of their own disadvantages. The ability to see President Johnson in his first public appearance since tragedy forced him into this office also meant that because of the noise in the place also made it hard to hear his words."

As the Navy ambulance carrying the late President and his widow makes its way through Washington, en route to Bethesda Naval Hospital, it passes Chesapeake & Potomac telephone trucks parked curbside. Workmen are below street level, installing miles of cables for weekend TV coverage.

Live from Dallas

While television and its audience attention on studio reports and coverage from Andrews Air Force base, an army of TV technicians have been working tirelessly at police headquarter in Dallas. Once President Johnson enters the White House, there will be nothing to report until Secretary of State Dean Rusk and others make their own return from Andrews. The focus returns to Dallas. Viewers soon become familiar with new faces and names: Chief of Police Jesse Curry, Dallas County District Attorney Henry Wade and Homicide Captain Will Fritz.

Power cables for cameras, lights and microphones and broadcast ephemera snake up the exterior of the Dallas Municipal Building, through windows and down the hall of the third floor offices. To provide for the demand of coast-to-coast television, the phone company had to clear lines. To connect a TV picture from WBAP to NBC, the phone company link ran from downtown Dallas to WBAP in Fort Worth, back to Dallas, from Dallas to Los Angeles, from Los Angeles to San Francisco, from San Francisco to Denver, from Denver to Oklahoma City, from Oklahoma City to Buffalo and from Buffalo to NBC in New York.

The police station hallway was elbow to elbow with reporters. Bulky TV cameras brought from the studios of local stations for live telecasts took up the most room. Every time a door opens is an invitation to reporters and lights, cameras, questions. Police Chief Jesse Curry and Dallas County District Attorney Henry Wade are each subjected to an impromptu press conference each time they enter the hall. The screen is full with reporters facing away from the camera. Chief Curry announces Oswald has been charged with the murder of the President. He has not confessed, nor has he made a statement. As District Attorney Wade attempts to answer a question, another reporter and; another interrupt. "Are you gonna bring him out?" Can you get a room where we can get a picture of him?" Can we get a press conference where he could stand against a wall?"

The Gun

The make of the rifle recovered is identified as a German Mauser, a .30 caliber rifle, a Japanese make and an Argentine make before the information is clear. The rifle is a 6.5 mm Mannlicher Carcano bolt action rifle with a 4x sight. Oswald ordered the rifle the previous March using a post office box he rented under the name Alek Hidell. Oswald also used the P.O. Box to order a revolver from another company.

When Lieutenant J.C. Day carries the rifle from the evidence lab and walks down the third floor, he holds it high above his head as not to give a clear look. The effect is a display of a grotesque trophy.

JC Day Friday evening via CBS

Back in the Studio

CBS produced an extended Evening News broadcast with Cronkite anchoring with reports from George Herman, Roger Mudd and Marvin Kalb in Washington. From Dallas, Dan Rather recaps.

George Herman outside the White House for CBS

From New Orleans, Bob Jones of WWL reports on Oswald's Fair Play for Cuba activity and; Bernard Eismann reporting from New York's Times Square. Harry Reasoner then hosts evening coverage, Dan Rather reports from KRLD with video of Oswald in custody. In answer to a reporter, Oswald declares "I did not shoot anyone."

NBC produced a ninety minute edition of the Huntley-Brinkley Report. Gabe Pressman reports from the deserted Times Square. Movie houses and theaters are closed; the great white way of Broadway is dark. Robert Goralski reports on the arrival of President Lyndon Johnson's arrival at the White House. As the night goes on, Bill Ryan narrates a Kennedy retrospective. From WBAP in Fort Worth, Robert MacNeil summarizes the events of the day. Nancy Dickerson reports on President Johnson. Defying Secret Service advice, President Johnson returns to his residence in the Spring Valley neighborhood of Washington, The Elms. Tom Pettit's first report from the Dallas Municipal

Building jail is by phone. Back at Andrews Air Force Base, Herb Kaplow reports live on the return of the cabinet. Secretary of State Dean Rusk makes a few remarks, from the same spot where President Johnson spoke hours before. Frank McGee reports Lee Harvey Oswald has been charged with the President's murder.

From Washington, David Brinkley summed up the day by "About all that could happen has happened. It is one of the ugliest days in American history. There is seldom any time to think anymore, and today there was none. In about four hours we had gone from President Kennedy in Dallas, alive, to back in Washington, dead and a new president in his place. There is really no more to say except it has been too much, too ugly and too fast."

NBC signed off at 1:02 a.m.

Background

The Civil Rights movement of the era rose in crescendo during the first week of May, 1963, in Birmingham, Alabama. When protestors would not disperse, public safety commissioner Bull Connor replied with violence. Television viewers, newspaper and magazine readers across the country were incredulous seeing filmed reports and photos of authorities wielding clubs, siccing police dogs and spraying fire hoses on demonstrators, many of them children. It also inspired some cities and their police departments and their businesses to become more media conscious. Southern and Southwestern cities did not wish to appear one-dimensional, to be painted with a broad brush of Birmingham.

One idea was to cooperate with the news media, be friendly.

More Background

On October 12, 1963, the Oklahoma Sooners were pasted 28-7 by the Texas Longhorns at Dallas' Cotton Bowl. When pre game partiers, fans of each team had too much to drink and got rowdy, the law moved in and made arrests. When a TV reporter filming got to close for one police officer, the cop raised his hand, blocking the camera lens. A few days later, the police chief was invited to the TV station where he was shown the film. After that, an edict was issued to officers: Do not interfere with the press.

On the night of Friday, November 22, a dilemma for the Dallas police was how to balance the needs of the of information hungry reporters and make clear to the assembled press (including TV cameras broadcasting live) that a professional police department was investigating the murders of President Kennedy and Dallas Police Officer Tippit. The police also had to balance the freedom of the press and the rights of the accused.

An issue presented more than once by reporters: was Oswald represented by counsel? Both Chief Curry and District Attorney Wade replied Oswald has been advised of his right to counsel. Representatives from the Dallas chapter of the American Civil Liberties Union visited the police station to assist Oswald and were rebuffed. An attorney could not be appointed by the court until Oswald appeared in court.

Live from Dallas, again

CBS provides America with its first look at Oswald in the wee hours of Saturday morning. Dallas police accommodated the press by making Lee Harvey Oswald (at least one law enforcement official wrongly said Oswald's middle name was Harold) available for the most bizarre photo opportunity ever. The authorities did not want any speculation the suspect was subject to mistreatment.

Oswald is brought to the assembly room where he faced reporters, cameras and microphones. "I don't know what this is all about, I didn't do anything. I know nothing more than that; I do request somebody come forward and give me legal assistance," he says. "Did you kill the President?" a reporter asks. "No, I've not been charged with that. In fact, no one has said that to me yet. The first thing I heard about that was when the newspaper reporters in the hall axed (sic) that question," Oswald answers. "You have been charged," the reporter responds. Disgust registers on Oswald's face. As police lead him away, another reporter asks how he got his swollen eye. Oswald responds, "A policeman hit me"

What Oswald did not say was the injury occurred as he resisted arrest.

Saturday in Washington

As Friday was a day of national shock, television Saturday offered a day for national mourning. Hugh Downs and Barbara Walters begin NBC's broadcasting at 7 a.m. with the Today Show. CBS follows an hour later with Mike Wallace anchoring from the newsroom. CBS' first report is from George Herman. A film crew had spent the night outside the White House to record the arrival of President Kennedy's casket at 4:30 a.m.

It was a grey, chilly, rainy day in Washington. Correspondents in Washington reported on the comings and goings at the White House and what they knew of funeral and burial plans. Reports feature dignitaries arriving at the White House. Former Presidents Eisenhower and Truman, prominent members of Congress and the Supreme Court all pay respects On occasion, the cameras turned to Lafayette Park where the ever present mourners braved the rain.

To remind the world Washington's government was stable, despite the blow of instability, the White House releases photographs of President Johnson as he meets with Eisenhower, Truman and Secretary of Defense Robert McNamara and Secretary of State Dean Rusk. The White House also announces LBJ will deliver an address to a joint session of Congress Wednesday at 12:30 pm to be carried live on television and radio.

Mrs. Kennedy allows TV cameras into the East Room for 15 minutes to record for news coverage. Information of President Kennedy's funeral comes out in dribs and drabs. The President's casket will be moved to the Capitol Rotunda for public visitation Sunday; Mrs. Kennedy will walk in the procession from the White House to Capitol. A Monday funeral Mass is to be at St. Matthew's Cathedral.

Saturday in Dallas

By Noon Saturday, over 400 reporters have gathered at the city jail in the Dallas Municipal Building, home of the police department and city jail. Reporters move as a group. TV lights go on and, cameras record anytime a door opens. What begins as Chief Jesse Curry stopping to answer a question or two turns into a ten minute press conference. Curry starts by announcing the Dallas police, the Sherriff's department, the Texas Rangers, the Secret Service and the FBI are all working the investigation in concert. Curry declares there is no evidence linking Oswald's Soviet past to a plot or conspiracy.

The three reporters directly in front of Curry are Bob Clark of ABC, KRLD's Bob Huffaker and Tom Pettit from NBC. About Oswald, Clark asks "Is there any doubt in your mind that this man killed the President?" Curry replies, "I think this is the man that killed the President." Huffaker asks if, before the President's visit, Oswald was subject to surveillance. "We in the police department did not know he was in Dallas. I understand the FBI did know he was in Dallas." Curry confirms a reporter's information; the handwriting on the order rifle order form from a Chicago sporting goods dealer matched Oswald.

Separately, Homicide Captain Will Fritz, Oswald's interrogator, tells NBC's Pettit "This is the man who killed Kennedy; we have a cinch case against him." This is too much for FBI Director J. Edgar Hoover, watching TV in Washington. He makes a call with a message to Fritz and Curry: Stop talking.

 The next time Curry stops to answer questions, he makes sure to take back his "understanding" the FBI knew Oswald was in Dallas.

All of this is unprecedented. City Manager Elgin Crull (who appointed Jesse Curry to Chief of Dallas Police in 1960) encouraged Curry to accommodate the media. No one wanted Dallas accused of running roughshod over the world press. This was an era before police departments had professional media advice, public information officers and crisis managers. The subsequent lesson learned by Dallas and its police department informed leaders of municipalities and industry the world over. It is one thing to accommodate reporters from local newspapers and TV; it is a different animal when the world media is at your door.

Reporters are told to back up to the wall, to clear a path and not to shout questions at the suspect as police Oswald escort to an interview room. "We may have to ask you to get back beyond this door," the officer warns. A question follows "In that event can the camera men be in front?"

Newspaper reporters come to a realization: each time there is a media scrum before Chief Curry or, any object of attention that weekend: TV gets the front row. So many out of town reporters with their cameras and microphones and cables and lights need room, demand attention, and move print reporters to the back of the pack. Sharp elbows are essential. Each reporter was mindful of colleague and competitor but, as a group, all that time scrambling for a morsel of information on the job and on their feet adds tension.

Everyone needs a nap.

Oswald emerges in a white T-shirt, flanked by Captain Fritz and a plain clothes officer. Questions travel and he responds "I have been answering the questions as fast as you people ask them. I emphatically deny these charges." As police lead him down the hall, Oswald passes NBC's Pettit he speaks directly into his microphone. "I wish to get in touch with Mr. John Abt in New York", he says.

When Oswald is out of reach, reporters turn to District Attorney Henry Wade. He will try the case himself he says, he will ask for the death penalty. In capital cases, he is 23 for 24. His matter of fact attitude is evident as he tells WFAA's Paul Good, "I figure we have enough evidence to convict him."

Oswald's wife Marina and his mother, Marguerite, are escorted down the hall to an elevator, to visit Oswald on the fifth floor. Reports follow them to the sliding doors. A reporter addresses Marina. The Soviet-born woman, who does not speak English, has learned one phrase. "Go away."

One Dallas TV viewer, unclear if Oswald understands his right to counsel, satisfies his curiosity by going to the county jail. He is H. Louis Nichols, president of the Dallas Bar Association. Chief Curry welcomes Nichols and takes him to Oswald. Oswald tells Nichols he is depending upon his family to contact John Abt, a nationally known attorney who has defended communist sympathizers. After meeting with Oswald, Nichols steps out of the elevator only to find the reporters waiting for him. Privately, Chief Curry tells Nichols he is glad he visited. All doubt as to whether Oswald knows his rights can be laid to rest.

As afternoon becomes evening and there is less and less news coming from the Dallas Municipal Building, TV turns its attention to Washington and, the New York studios. In Dallas, everyone is cranky and wants to know when Dallas police will move Oswald to the county jail. Most reporters have been there for 24 hours, some, closer to 32. There is no set time but, Chief Curry finally says "If you get there by ten, I don't think you'll miss anything."

One idea was to avoid a media crush by moving Oswald overnight. To have done so, the Chief reckoned, would leave his department open to charges Dallas police had something to hide and he knew all eyes were on Dallas. A morning transfer, for the media to chronicle, became consensus.

TV and Mass Communication Theory

In his 1922 work Public Opinion, Walter Lippman posited that mass media had become the prism through which people see the world. Building on that theory, Bernard Cohen concluded the press does not tell the public what to think but; the press succeeds at what the public think about. This was crystallized by Maxwell McCombs and Donald L. Shaw in their study of the 1968 election with the Agenda Setting Theory. Simply said, the frequency of news reports on a topic correlates with how the public regards the urgency or importance of the story. Perhaps Harry Reasoner had Lippman in mind when he closed CBS's broadcast Saturday night.

"At the end of this second day of concentrated national grief and one event, it may be time to stop and think about our own attitude. Introspection is proper in sorrow as it is at any time because mourning if it becomes a fixed and purposeless moan about the cruelty of fate can be habit forming. In Norwalk, Ohio, today, a fire burned up a home for the elderly and, about 63 old men and women died. There is a way of thinking about our knowledge of God that might make you say that, in his sight, that event was 63 times as important as the death in Dallas. In the national attention, those 63 have scarcely had a place. They get six inches of type in the Sunday edition of the New York Daily News, for instance, just above a little item about a man who stole some money from a department store.

You might think that we are out of proportion; that the national dirge that fills these days is inappropriate. Either we should do more, mourn all the time for everybody, or maybe do less. There were for instance, some calls last night, to CBS in New York, from citizens complaining about missing their normal Friday night programs. Our operators, I understand, were polite. We are not out of proportion. We are not dishonoring the 63 old folks or the thousands of others who died yesterday and today and will die tonight and tomorrow. We are not God. We are a nation of men who tempt, with honor and reward, all kinds of men to serve us. When one is especially worthy, especially important to us, and becomes a sacrifice as well as a leader, it is entirely appropriate that we do him great honor. We are all dying. And what we feel about John Kennedy is not so much sadness that he met his appointment a little sooner, but a gratitude and love for a man who would make that appointment for us. There is only one reservation; it must not be a habit. When President Kennedy announced the quarantine of Cuba, one reporter suggested what he wanted from his country was intelligent support, not intoxicated belligerence. It seems likely that what this man would want from his martyrdom would be a considered dedication, not a pointless self-pity."

"The news today; Monday will be the day of the funeral and a national day of mourning. Tomorrow the President's body will be moved in solemn style from the White House to the Capitol to lie in state. Beginning at 9.am. Eastern Time tomorrow, CBS News will begin continuous coverage of the day's events as we have and as we will.

This is Harry Reasoner. Goodnight."

Sunday morning

Television news Sunday morning is as subdued and hushed as a ceremony at a house of worship. The President's casket is to be moved from the White House to the Capitol Rotunda for public visitation until 9 p.m.

Sunday in Dallas

Reporters, photographers, and television technicians, with their heavy cameras on tripods, cables and, portable studio lights set up shop in the basement of the Dallas Municipal Building to televise the transfer of Lee Harvey Oswald. An armored car sits at the garage entrance. Too big to back in, the armored car is to serve as a decoy and block the entrance. As the armored car diverts attention, Oswald would be slipped into an unmarked police sedan with an escort for the ten block ride and delivery to the county facility.

Any law enforcement official, detective or uniformed cop had learned to deal with the crowd of reporters this weekend but, Sunday morning was an eye-opener. Newspaper reporters, photographers, TV and radio reporters and their equipment packed the basement garage of Dallas Police headquarters. There were reporters of Dallas and other Texas newspapers, the dailies from New York, Atlanta, New Orleans, Chicago and Los Angeles. The wire services were feeding newspapers and radio the world over. The mobile units for CBS and NBC parked on the street. ABC decided to forgo the departure scene and established its own broadcast beachhead outside the county jail.

The mobile unit for NBC was less a marriage of technology of TV and automotive knowhow than necessity. After breaking down Friday, the NBC remote was ferried around town by tow truck. Whatever the look, it successfully relayed pictures and sound from reporter Tom Pettit to NBC New York.

The CBS unit parked on the street was a bit roomier. Reporting from inside the vehicle was network veteran Nelson Benton. From the van, Benton could rely on TV monitors to provide multiple views. He'd know which camera the home viewer was seeing, the garage or exterior. Reporting from the garage floor was Bob Huffaker of KRLD, the Dallas CBS station. With his headset and TV monitors, Benton could hear the CBS control room in New York. Huffaker had no headset; he would rely on hand signals delivered by his floor director. Huffaker was the right reporter for this assignment; he had covered police and the courts and knew many cops by name.

Back in the garage, the police got to the business of setting boundaries. TV cameras and some photographers stood behind a rail opposite from where Oswald would emerge. Reporters lined the wall. For NBC, the plan was to carry the transfer live, switching to the scene on Tom Pettit's cue. CBS would record the Oswald transfer, opting to stay with the live coverage from Washington, where Roger Mudd was reporting from the Capitol steps, explaining the procession of the President to the Rotunda. After an essay by Harry Reasoner in New York, CBS would switch to Dallas.

There's a Shot

As the moment drew closer, Pettit called to New York "Give me air! Give me air!" A New York director spoke into his anchorman's earpiece and, Frank McGee made a flawless toss to Pettit.

At 11:21 a.m. Oswald emerged from a corridor into the garage, handcuffed to Detective Jim Lavelle. Oswald and Lavelle squint at the bright lights. "There is Lee Oswald," Pettit began to narrate. A form emerged from the crowd of reporters with the sickening sound of a gun blast. A scuffle results as Pettit continues. "He's been shot! He's been shot! Lee Oswald has been shot. There's a man with a gun. There's absolute panic," Pettit continued, his voice relaying alarm.

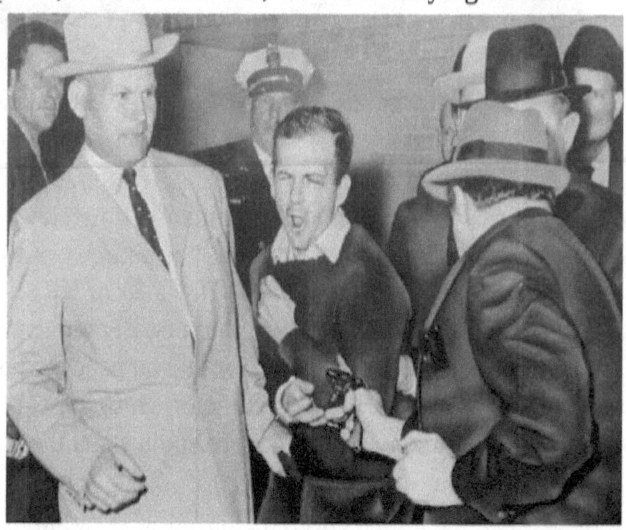

The resulting tussle of detectives and cops for control of the gun and the gunman obscured the view. Lee Harvey Oswald is lost in the tangle. After police retreat to get Oswald first aid, Pettit turned to witnesses, fellow reporters Frank Johnson of United Press International and Geoff Edwards, of KHJ Radio of Los Angeles.

CBS dumped out of Roger Mudd's report in midsentence to Harry Reasoner at the anchor desk in the CBS newsroom. Looking stunned just for a moment, Reasoner announced "We are now switching to Dallas where they are about to move Lee Oswald and there's a scuffle in the police station." "Oswald has been shot," an off-camera Huffaker exclaimed. In the melee, people were stepping on his microphone cord and nearly, Huffaker himself. Huffaker positioned himself in a bit of a crouch before the camera. Over his right shoulder, Tom Pettit continued his report for NBC, over his left shoulder was a mess of reporters. The crowd parts enough to show an ambulance pulling into the garage.

Police link arms to keep reporters back. The view is blocked. Kneeling off-screen, Huffaker tapped Sergeant Pat Dean of the Dallas Police Department on the hip. "Pat, can ya move? Pat, move a little, just to your left." Now on a stretcher, Oswald is rolled out from behind closed doors to the ambulance, his left hand drags on the floor. Huffaker continues: "Here is Oswald now; he is ashen and unconscious as he is now being moved in. He's not moving. He's in the ambulance now…"

Two detectives climb in after Oswald and the ambulance tailgate goes up. CBS switches to a camera outside as the armored car moves from the mouth of the garage. The ambulance follows, its extra weight making it a low rider. CBS switches back to Harry Reasoner who explains the video tape has been "re-racked" (TV lingo for cued). "We will now roll it, and you will see it as it happened."

Once the ambulance is gone and Oswald's shooter is in custody, reporters interview reporters. Two reporters with the best view are Francois Pelou of Agence France-Presse and WNEW New York radio's Ike Pappas. Both saw a man in a dark hat squeeze through and saw the flash from the muzzle.

Reporters crowd as Huffaker interviews Officer Pat Dean. Most reporters deferred to Huffaker, for he was the local of the bunch. One or two tried to barge in with a question of their own but others let Huffaker lead. Huffaker and Dean are each walking a fine line. Dean was not speaking for the Police Department officially, but responding to the questions by a reporter he knew.

Yes, police saw the man who shot Oswald, yes, they know his identity, yes, they knew what business he was in, and yes, so far as Dean knew, the gunman is undergoing questions upstairs. More than that, he would be reluctant to say. Afterward, NBC's Pettit said to Huffaker "I appreciate what you just did."

Roger Sharp was reporting for ABC outside the county jail as the Oswald shooting is announced and spectators who came for a firsthand look let out a whoop. Whatever Sharp might have anticipated, whatever he rehearsed for his live shot became notions he had to dismiss. Improvising, Sharp interviewed those who showed up out of curiosity.

At the Capitol, CBS' Roger Mudd takes notice as the news travels through the crowd; many heard the story over transistor radio.

To the Capitol Rotunda

Sunday is a bright, crisp fall day. President Kennedy's casket is carried from the North Portico of the White House to a caisson, to be transported by a team of six horses. Nancy Dickerson speaks only to identify mourners to the television audience. Eighteen military drummers, theirs, the only sounds to be heard, follow. Ten cars, with the President's family and government officials, follow to Capitol Hill. It is estimated 300,000 line Pennsylvania Avenue to watch the procession.

At the Capitol, the US Navy band plays "Hail to the Chief." Pallbearers of each branch of the military carry the casket up the Capitol steps to the Rotunda. Speaker of the House John McCormack of Massachusetts, Chief Justice Earl Warren and Senator Mike Mansfield of Montana deliver eulogies.

ABC faced a dilemma when the network learned WFAA's film of the Oswald shooting was processed. The WFAA footage was the best angle yet. ABC also had news on Oswald's condition. However, the camera was fixed on the casket, to break away to air the shooting, ABC considered, would be in poor taste. The solution was to switch to a camera outside as President Johnson's limousine drove up. For ten seconds, ABC superimposed the news at the bottom of the screen "Oswald is Dead". The film could wait.

After President Johnson delivers a wreath, Jackie leads daughter Caroline to the casket. They kneel and Mother and daughter kiss the flag.

As the procession leaves the Capitol, Frank McGee announces on NBC that Oswald has died.

Parkland, again

For the second time in 48 hours, Parkland Hospital is besieged with reporters after a shooting. From the same classroom where Malcolm Kilduff announced JFK's death, Parkland administrator Steve Landregan informs reporters of every step to save Oswald. Following a premature report Oswald has died, Landregan makes it a point to tell reporters "It's impossible for anyone outside this room to find out before you do."

A few minutes later, Doctor Tom Shires enters the room. Reporters gather as Landregan pleads to them to allow Dr. Shires to make a statement. The questions come anyway. "He died at 13 …1:07 p.m., of his gunshot wound…that he received." Out of his element, Shires asks generally, "Would you like to ask questions or what?" Shires reports the bullet entered Oswald's abdomen on the left, passed through the spleen, pancreas, the aorta, vena cava, the right kidney and the liver before coming to rest just below the skin on his right side. The severe blood loss caused Oswald a cardiac arrest. Effectively, he bled to death.

Landregan's vow was true. The press was notified before Oswald's family.

Later, both Dan Rather of CBS and NBC's Frank McGee review video of the Oswald shooting. The video is slowed, reversed, examined and reviewed in a manner to which sports fans will grow accustomed. Rather also narrates film shot by KRLD's George Phenix. Phenix was behind the railing in the garage and kept recording after he was jostled. He worried he missed the action. He got a great angle.

Police identify Oswald's assailant as Jack Ruby aka Rubinstein, the owner of two Dallas nightclubs who fancies himself a police insider. His name rings a bell with some reporters. Ike Pappas reaches into his pocket and finds a business card for the Carousel Club. Printed on the card is "Your host, Jack Ruby." Ruby was the sort to hang around the police station, show up at a fire or a trial. This weekend, he has been a constant presence in the Dallas Municipal building, giving out-of- Towner's the lay of the land, providing some reporters with soft drinks and sandwiches and, handing out cards to his strip club. When recognized at a press conference Saturday, Ruby claimed he was assisting Israeli press.

Until this weekend, Ruby has been nothing more than a well-meaning pest.

Sunday Evening

The networks air panel shows, retrospectives and, a concert from the Los Angeles Memorial Coliseum.

The Capitol

The public pays respect by filing past each side of the casket. They come from all walks of life, all ages and races civilian and military. They are wearing their Sunday best, high school letterman's jacket, casual dress or, work uniform. A recurrent observation by reporters on each network is the mourners under age 21.

The line to walk past President Kennedy's casket is more than forty blocks and plans to close the Rotunda doors by 9 p.m. reconsidered. The new cutoff is announced, 8 a.m.

Mourners stood patiently, their numbers growing. An emotional image is of the Capitol Dome, aglow against dark Washington night. As the camera pulls back, viewers at home get an idea how long the line is. At 8 p.m., Capitol Hill police make way for a special visitor. Mrs. Kennedy walks slowly to the bier, kneels and has a private moment.

By midnight, it is estimated over 100 thousand had paid their respects. The line is still miles long. Television's overnight presence was constant, subtle and unobtrusive. Commentators seldom spoke and, when they did, they spoke in hushed tones. One instance was to note former heavyweight boxing champion Jersey Joe Walcott was among the mourners just after 3 a.m.

Conducting a live report for the Today Show at 7:25 a.m., Jack Lescoulie did no interviews as "it would not be proper to stop them as they walk by."

Heads of State flew into Washington from all over the world. Television crews broadcast their arrivals at Dulles Airport, interviewing few. Among the dignitaries: Lester Pearson, Canadian Prime Minister, Prince Philip and Sir Alec Douglas-Home of Britain, Eamon de Valera, President of Ireland, French President Charles de Gaulle, Princess Beatrix of the Netherlands, King Baudoin of Belgium, Swedish Minister Olaf Palme, Heinrech Lubke, President of West Germany, Mayor Willy Brandt of West Berlin, Queen Frederika of Greece, Prince Akihito and Premier Hayato Ikeda of Japan, General Tran Van Don of South Vietnam, Emperor Haile Selassie of Ethiopia and, Anastas Mikoyan, first Deputy Premier of the Soviet Union.

Monday

Three months earlier, the networks had pooled their resources in televising the March on Washington for Jobs and Freedom. The television coverage of The March, weeks in planning, served as a template for JFK's funeral.

Workers for Potomac Electric (Pepco) and C&P telephone toil around the clock to provide power for the TV units and eleven miles of video and audio cables. Each network contributed cameras and personnel for pool coverage. ABC provided 138 employees; NBC dispatched sixty engineers. Helming the network coverage was CBS producer Art Kane, who produced the March on Washington. Kane supervised 41 cameras in 22 locations from the White House, along the procession of the cortege to the Capitol Rotunda where President Kennedy would lie in state overnight. Directing the telecast was Norman Gorn.

Cameras, cables, microphones and other equipment are trucked in from television stations from New York, Baltimore, Philadelphia, Richmond and, New Haven, where they were to be used for the Harvard-Yale football game. Long-range camera lenses are sent from Japan; many cameras are active only a moment before delivering a picture to 175 million viewers. Carpenters built platforms for cameras and reporters for the telecasts. One platform was too close, for the comfort of police. The platform was disassembled, moved back, and rebuilt. The effort proved to be for naught when the new angle made the camera useless.

A Worldwide Television Audience

The Honor Guard pauses as it carries the casket from the Rotunda; a Marine band plays the Hail to the Chief. It proceeds slowly down the steps and places the casket on the caisson to the hymnal "O God of loveliness". The procession moves to the White House. It is from there Mrs. Kennedy insists on following the funeral cortege to St. Matthew's Church on foot. Flanking her are the late President's brothers, Attorney General Robert Kennedy and Senator Ted Kennedy of Massachusetts. Other members of President Kennedy's family and visiting heads of state followed. President Johnson, Mrs. Johnson and daughters Lucy and Linda walk also, much to the concern of the Secret Service.

The riderless horse, a tradition dating from centuries past, follows the caisson. Black Jack's handler has all he can do to keep the steed in line. He is as Nancy Dickerson later described, "full of beans". There are moments when the beating drums and horse's hooves are all that can be heard.

Over a million people lined the streets from the Capitol to the White House to St. Matthew's. Mixed in were 4000 Pentagon and District of Columbia police officers. The State Department assigned 250 officers to guard the foreign dignitaries – 10 for de Gaulle alone. Forty FBI and 64 CIA agents secured the motorcade. NBC televised the procession to Saint Matthews to twenty- three countries, including the Soviet Union and Japan. The transmission was via Relay, the satellite of NBC parent company RCA. Commentary during the procession was minimal. Edwin Newman informed the audience the Navy band played Onward, Christian Soldiers as it passed the intersection of 6th Street and Pennsylvania Avenue. Chopin's Funeral March is the next piece.

Richard Cardinal Cushing of Boston meets Mrs. Kennedy at the steps of Saint Matthews. David Brinkley notes Cardinal Cushing, who will celebrate the funeral Mass, married the Kennedys ten years earlier. Mrs. Kennedy walks into the church holding hands with her children, Caroline, who will be six Thursday and, John, who is three today. During the service, NBC has Father Leonard Hurley serve as an analyst and translator when Cushing speaks Latin.

After the Mass, as television focuses on the Kennedy family gathering at the front steps of the church, the Honor Guard carries the caisson. Mrs. Kennedy is holding hands with daughter Caroline on her right and her son on her left. Mother takes a paper from the boy's hand and whispers. John John turns and offers a salute to his father. The image captured by United Press photographer Sam Stearns endures.

Television provides a panoramic view of Washington as the procession makes its way to Arlington National Cemetery. David Brinkley gently intones Mrs. Kennedy has requested an eternal flame. Herb Kaplow sparingly narrates the graveside ceremony. NBC sent word for commentary to be brief and minimal.

On occasion, a reporter for CBS talks too long, his microphone is cut off.

The Marine Band played The Star Spangled Banner before the Honor Guard removes the casket from the caisson. The Air Force bagpipes play "Mist Covered Mountain". As the mourners gather at graveside, fifty jets, thirty Air Force F-105 and twenty Navy F-4B aircraft, flyover Arlington in salute. The roaring engines of Air Force One follow; Air Force Pilot Colonel James Swindal dips the right wing of the plane in honor. The Irish Guard performs its drill and then, the foreign leaders assemble. Conspicuous by his stature is President de Gaulle (he stood six foot five), particularly next to Emperor Haile Selassie, a foot shorter.

After Cardinal Cushing performs the graveside ceremony, the 3rd infantry fires a twenty-one gun salute. Army Sergeant Keith Clark plays Taps. The Marine Band plays a hymn as the Honor Guard folds the flag.

An Officer hands Mrs. Kennedy a taper so she may light the eternal flame.

Mrs. Kennedy is handed the flag that draped her husband's coffin. The ceremony was over and, the Kennedy family leaves.

Television is the guest that stays too long. John Metzler, the superintendent of the Arlington National Cemetery notices the television camera still staring at the grave; he approaches the director in the TV truck. When the director explains New York resists his cue to switch away, Metzler gives an order to his crew. "Pull the plug!"

Viewers have no idea. The next image onscreen is of United Nations Secretary General U Thant, Doctor Ralph Bunche, Hayato Ikeda, Anastas Mikoyan and diplomats from South Korea, Poland, Finland, Tunisia and Somalia entering the Diplomatic Reception room of the White House. Only then does Merrill Mueller, Ed Newman, and Nancy Dickerson, all of NBC, speak above a whisper as they identify the dignitaries and comment.

Dallas TV stations switch to live coverage of the funeral for Officer J.D. More than 700 police officers from Texas and nearby states attend.

Lee Harvey Oswald is buried at Rose Hill Cemetery in Fort Worth. Reporters step in, for lack of pallbearers. There is no live television.

Afterward

News coverage of that weekend across the globe was sensitive and often, reverential. Only China, East Germany and Poland aired no television news of Kennedy's assassination. TASS, the official broadcaster for the Soviet Union, was critical of US law enforcement and news coverage in describing Oswald as a communist. The New York Journal-American is singled out. TASS did not distinguish between news reports and editorials.

Television's achievement came at a high price. The three TV networks cancelled all entertainment programs and commercial advertising for the weekend at an estimated $40 million loss. Equipment and personnel cost another $3 million to televise weekend ceremonies and funeral Monday. By Sunday morning, some mercenary local stations began to slip in a few ads. For the record, the distinction of capturing the largest audience is NBC's then, CBS and ABC. It is worth noting, at the time, the Huntley-Brinkley had more viewers than the CBS Evening News with Walter Cronkite. By 1967, the CBS Evening News surpassed Huntley-Brinkley. Four of ten Americans were watching as Jack Ruby shot Oswald. An hour later, eight of ten Americans had seen it. At its highest levels, Friday, Sunday after the Oswald shooting and, during President Kennedy's funeral, 93% of Americans watched the three networks.

The saturation coverage of those four days in November, 1963, was television's first opportunity to fulfill television's potential and power as a technological breakthrough and uniting presence of influence and enlightenment.

The assassination as recorded by Abraham Zapruder film was not broadcast until March, 1975, when televised on ABC's AM America (retitled Good Morning America later that year).

Bibliography

In addition, there are many videos of the Kennedy assassination coverage online.

The webpage "JFK's Assassination as it happened" was particularly helpful. It can be

found at http://jfk-assassination-as-it-happened.blogspot.com/.

There are also news excerpts on David Von Pein's JFK Channel on YouTube.

An extended clip of the Oswald shooting is on an NBC website http://www.nbcuniversalarchives.com/nbcuni/clip/51A02392_s01.do

The Archive of American Television has an interview with WRC director Max Schindler

at http://www.emmytvlegends.org/interviews/people/max-schindler

Encyclopedia of Television News

Edited by Michael D. Murray

Oryx Press, 1999

Alan Schroder

Pages 119-120 print reporters resent TV news conferences JFK first announcements, Oswald shooting 4/10 and 8/10

John F. Kennedy

Remarks at the Breakfast of the Fort Worth Chamber of Commerce.

November 22, 1963

http://www.presidency.ucsb.edu/ws/?pid=9538

JFK's Last Hundred Days

The Transformation of a Man and the Emergence of a Great President

By Thurston Clarke

Page 213 Hats.

When The News went Live

Dallas 1963

Bob Huffaker, Bill Mercer, George Phenix and Wes Wise

Taylor Trade Publishers 2004

Page 116 Adlai Stephenson

Texas: A City Disgraced

Time Magazine

Friday, November 1, 1963

http://content.time.com/time/magazine/article/0,9171,875296,00.html

Cora Frederickson

E-mail: Gary Mack

Video Curator

The Sixth Floor Museum at Dealey Plaza

411 Elm Street

Dallas, Texas 75202 List of stations and reporters of TV pool coverage for Forth Worth Breakfast, Love Field reception and, Trade Mart lunch.

E-mail: Andra Bennett

Fort Worth Chamber of Congress

777 Taylor St #900

Fort Worth, TX 76102 Identity confirmation of Forth Worth Chamber of Commerce breakfast emcee.

Death of a President

William Manchester

Galahad Books

1967

Page 129-Dave Powers quote "Mr. and Mrs. America"

Page 149-365 police officers guarding at JFK's Love Field arrival

Four days in November: the assassination of President John F. Kennedy

Vincent Bugliosi

W.W. Norton & Co.,

2007

Page 40 Estimated reception crowd at Love Field

Nation: La Presidente

Time Magazine

Friday, June 9, 1961

http://content.time.com/time/magazine/article/0,9171,938093,00.html#ixz

JFK's remarks in Paris

President Kennedy Has Been Shot

A Moment-to-Moment Account of the Four Days That Changed America

Susan Bennett, Cathy Trost

The Newseum

Sourcebooks

2004

Page 15 Jackie's suit in Dallas is **NOT** pink.

Page 30 Merriman Smith, first UPI report

Ten Bells Signaled Moment in History

By Alan Robinson

The Associated Press

Thursday, November 17, 1988 NBC Correspondent Ryan alerted.

The Day Kennedy Was Shot: An Hour-by-Hour Account of What Really Happened on November 22, 1963

Jim Bishop New York

Funk & Wagnalls 1968

Page 198 David Brinkley saw Smith's UPI bulletin

Newsman, Mr. Peppermint reflect on JFK 45 years later

By Jason Whitely

WFAA-TV

Posted on August 15, 2009

http://www.wfaa.com/news/local/64518552.html

The Kennedy Assassination--24 Hours After: Lyndon B. Johnson's Pivotal First Day as President

By Steven M. Gillon

Basic Books

The Kennedy assassination and the American public: social communication in crisis.

Bradley S. Greenberg [and] Edwin B. Parker, editors

Stanford University Press, 1965

The Television Story in Dallas

Tom Pettit

Pages 61-63 Pettit dispatched from Burbank, Vanocur flies onto DC, TV feed to NY

Elmer W. Lower

Page 69 Howard K. Smith

Page 70

Tube of Plenty

Erik Barnouw

Oxford Press

1982

Page 331 TV cables at Dallas PD.

Four days in November: the assassination of President John F. Kennedy

Vincent Bugliosi

W.W. Norton & Co.,

2007

Page 604 Brinkley goodnights for NBC

Nation: Dogs, Kids & Clubs

Time Magazine

Friday, May. 10, 1963

http://content.time.com/time/magazine/article/0,9171,830260,00.html

When the News Went Live: Dallas 1963
Robert Huffaker, Bill Mercer, George Phenix, Wes Wise
Taylor Trade Publishing
2004
Page 46 Oklahoma-Texas game
Page 47 Elgin Crull
President Kennedy Has Been Shot
A Moment-to-Moment Account of the Four Days That Changed America
Susan Bennett, Cathy Trost
The Newseum
Sourcebooks
2004
Page 183 John Abt
Four days in November: the assassination of President John F. Kennedy
Vincent Bugliosi
W.W. Norton & Co.,
2007
Page 370-74 H. Louis Nichols
First Person
A Small Town's Grief 25 Years Ago
By Jim Walters
The Los Angeles Times
November 23, 1988
http://articles.latimes.com/1988-11-23/news/vw-470_1_years-ago
Seventy Hours and Thirty Minutes
NBC News
Random House
1966
Page 104 concert from Los Angeles

Page 124 Jack Lescoulie

Death of a President

William Manchester

Galahad Books

1967

604 Metzger pulls plug on funeral broadcast

606 White House reception following Arlington

Mikoyan Flies to Washington As Russians Praise Kennedy: Moscow...

By Henry Tanner

The New York Times

Monday, November 25, 1963

pg. 7

Television Pools Camera Coverage: Measures Set a Record for Distance

By Richard F. Shepard

Tuesday, November 26, 1963

The New York Times

Page 11

As 175 Million American Watched....

Newsweek

December 9, 1963

Pages 88-90 Monday funeral coverage, Art Kane, CBS microphones

Slain Policeman Is Honored by Dallas: Streets Are Quiet Oswald Buried Return to Normal Urged

By John Herbers

The New York Times

Tuesday, November 26, 1963

pg. 15

Death of a President

William Manchester

Galahad Books

1967

Page 530 Television audience

There are thousands of books on John F. Kennedy. The Library of Congress online catalog lists over 700 books in its collection on the John F. Kennedy assassination alone.

Photo Credits

President Kennedy takes a question at a press conference, November 20, 1962 Photograph by Abbie Rowe/ John F. Kennedy Presidential Library and Museum, Boston.

JFK, Caroline and John-John in the Oval Office, October 10, 1962
Photo by Cecil Stoughton
http://www.maryferrell.org/photos.html?set=JFKL-KENFAM
The President and Mrs. Kennedy, May 24, 1961
Photo by Cecil Stoughton
http://www.maryferrell.org/archive/photos/JFKL-KENFAM/Photo_jfkl-01_0093-ST125-4-61.jpg

President Kennedy speaks at a public rally outside the Texas Hotel before the Fort Worth Chamber of Commerce breakfast. Behind him are left to right, unidentified, Senator Ralph W. Yarborough, Governor John Connally, Vice President Johnson. Photograph by Cecil Stoughton/John F. Kennedy Presidential Library and Museum, Boston.

President Kennedy accepting handshakes after his remarks at the Texas Hotel parking lot. Photograph by Cecil Stoughton/John F. Kennedy Presidential Library and Museum, Boston.

CBS Evening News Anchorman interviews President Kennedy, Labor Day weekend, 1963. Photograph by Cecil Stoughton/ John F. Kennedy Presidential Library and Museum, Boston.

Mrs. Kennedy and the President arrive at Love Field, Dallas. "Mr. and Mrs. America", Dave Powers called them. Photograph by Cecil Stoughton/John F. Kennedy Presidential Library and Museum, Boston.

The CBS Bulletin slide.

Frank McGee, Chet Huntley and Bill Ryan report from 5HN, the NBC Flash studio.

ABC Anchorman Ron Cochran.

Walter Cronkite reports President Kennedy is dead.

Lee Harvey Oswald: National Archives/President John F. Kennedy Assassination Records Collection

Vice President Lyndon Johnson takes the oath of office flanked by Mrs. Johnson and Mrs. Kennedy. Administering the oath is Judge Sarah Hughes. Photograph by Cecil Stoughton/ John F. Kennedy Presidential Library and Museum, Boston.

Air Force one arrives at Edwards Air Force Base, November 22, 1963. Note the shadows caused by TV lights. John F. Kennedy Presidential Library and Museum, Boston.

Lee Harvey Oswald faces the media.

A WBAP camera moves into place before Oswald is brought out for transfer.

Tom Pettit reports from the garage of the Dallas Municipal Building.

Oswald is shot by Dallas night club owner Jack Ruby in the basement garage of the Dallas Municipal Building. Photo by Robert H. Jackson Dallas Times-Herald Dallas Municipal Archives, Dallas, Texas

Mourners wait at the Capitol Photograph by Abbie Rowe/John F. Kennedy Presidential Library and Museum, Boston.

The Honor Guard/Photograph by Abbie Rowe/John F. Kennedy Presidential Library and Museum, Boston.